A COMPREHENSIVE PREPARING EXHIBI HONEY SHOW

David Shannon

Edited by John Phipps

NORTHERN BEE BOOKS

A COMPREHENSIVE GUIDE TO PREPARING EXHIBITS FOR A HONEY SHOW

David Shannon

All rights reserved. No part of this publication may be reproduced, stored in a retrieval system, transmitted in any form or by any means electronic, mechanical, including photocopying, recording or otherwise without prior consent of the copyright holders.

ISBN: 978-1-908904-81-2

Published by Northern Bee Books, 2015
Scout Bottom Farm
Mytholmroyd
Hebden Bridge HX7 5JS (UK)

www.northernbeebooks.co.uk
Tel: 01422 882751

Printed by Lightning Source, UK

A COMPREHENSIVE GUIDE TO PREPARING EXHIBITS FOR A HONEY SHOW

.... and top quality hive products for sale!

David Shannon

YBKA Chief Honey Steward

Edited by John Phipps

NORTHERN BEE BOOKS

Cover image: Winning display at the National Honey Show, London. (J Phipps)
Back cover: The author - David Shannon

CONTENTS

Foreword - John Phipps 7

Introduction 9

1   Essential Equipment 11

2   The Preparation of Liquid Honey for the Show Bench 13

3   Two Methods of Preparing Creamed Honey 19

4   Cut Comb to Hive to Container 23

5   Sections 31

6   Cakes and Confectionary 37

7   Meads, Dry and Sweet; & Fruit Melomels 43

8   Observation and Nucleus Hives 51

9   The Display Class 55

10 The Preparation of Beeswax 59

Exhibits of honey, wax and cakes ready for judging. (D Shannon)

FOREWORD

I have kept bees for nearly forty-five years and not once exhibited any item in a honey show - either at local or national level. In fact, to be really honest, having attended many honey shows in the UK and elsewhere, I have been intimidated by the absolute quality of the hive products on display. In my eyes the hundreds of jars of honey in different shades; the excellent examples of comb honey with beautiful white capping; the perfect pieces of wax in block, candle or model form; the rows of meads; mouth-watering honey confectionary and cakes;— and yes, even observation hives showing colonies in miniature but in exceptionally good form; made me realise how much care and expertise had been carried out by the beekeepers to achieve such high standards. I also thought that the task of the judges must have been extremely difficult in order to award prize winning certificates in the face of such strong competition.

Honey shows have been popular at national and local level far in excess of a century - and those who have returned home over the years with the coveted prizes can take pride in the fact that they have achieved the highest recognition for their products. Of course, there is always the excitement too, for any competition creates amongst those taking part, the nagging thought that other exhibits are better than their own - even more so perhaps when they see what other beekeepers have entered. Is Mr or Mrs X going to sweep the board again - as a beginner do I have the remotest chance of beating those who have been in the game so long?

Well, here in Dave Shannon's book, no matter your age or experience in beekeeping, are clear instructions how to prepare any exhibit worthy of a prize. But following his procedures with care and patience, you will most certainly be on the road to success. In past years, given such information, I would no doubt risen to such a challenge.

John Phipps, Greece,
September 2015.

INTRODUCTION

I was asked by the editor of Halifax BKA Newsletter to explain, in a series, the processes involved in the preparation of honey and wax for the show bench. The articles were aimed mainly at beginner beekeepers and members who would like to get involved in honey and beeswax showing but are not quite sure of the process to get to the required standard. I went through the whole procedure from how to filter and prepare honey, the correct jars for the show, plus the labelling. I included all the various types of honey from the more popular light and medium honeys through to the process of creaming honey and the preparation of heather honey - both liquid and cut comb. The same applied to beeswax: how to get the best wax for showing; how to prepare it for a moulded wax show block; and how to make to wax sculptures and candles. By following the instructions readers should be able to place their exhibits in a local show and then the YBKA autumn honey show held at the Show Ground, Harrogate, (Countryside Live) and win prizes. Having achieved this readers will not only have the great satisfaction of gaining rosettes and prize certificates, but will also find that with these items displayed amongst their products on a stall will enhance your honey and wax sales to friends and customers alike.

It stands to reason that in order to cover all the various types of honey displayed on the show bench, more than just one colony of bees will be needed so that they can be moved to cover several different crops and nectar types. However, excellent honey can be produced from only one hive in the garden - if it is processed correctly.

In whichever field the beekeeper covers in connection with honey shows, there is nothing that gives more pleasure than beating fellow competitors and having the satisfaction of gaining a reputation for producing a top quality product.

I hope the methods described in this book will allow beekeepers to achieve that goal.

David Shannon, September 2015.

CHAPTER 1
Essential Equipment

Here is a list of the things you will need to either make, buy or borrow to achieve good results. Winter is a good time of year for beekeepers to make or purchase those essential things needed for the forthcoming season whilst the bees are dormant.

Honey

- A warming cabinet - preferably your own (either make or buy one). Beekeeping books and internet provide good diagrams and instructions on how to make one and its quite easy. Preferably with a thermostat ranging from 0 to 45 degrees centigrade.
- A plastic or metal fine strainer - for getting the larger pieces from your honey when extracting it.
- A very fine fabric filter - these can be obtained at Thornes or other beekeeping equipment stockists. They are not overly expensive.
- A honey creamer
- Access to a refractometer - either your own or a friends, but essential to check your honey's water content.
- Some 1lb jars with gold screw top lids to match - as most shows, but not all, insist on these as standard honey jars.
- A good LED torch - for looking at and checking your honey when filtered for any crystallisation or minute detritus or hairs.
- A magnifying glass - for the same use as above
- Access to a set of official grading glasses - to check the class your honey should be in prior to the show.

Wax

- A large old saucepan
- A Ban-Marie: wax melting.
- Pyrex glass dish (for a wax mould).
- Roll of clinical lint, available from boots.

CHAPTER 2
The Preparation of Liquid Honey for the Show Bench

Honey classes are open for dark, medium, light, liquid and set exhibits.

(D Shannon)

After keeping bees for two or three years successfully, and achieving a surplus of honey from your hive or hives, embarking on showing your honey at your association or local county honey show comes as a natural progression. For the absolute novice beekeepers among you it is essential to first master the art of tending your bees and extracting your honey successfully. In this book I am hoping to take you from being a good beekeeper to being a good exhibitor capable of entering your honey in any show and having a very good chance of winning a prize in your chosen show class, irrespective of the number of hives you own. The key to producing a good liquid honey is care and attention in every detail of your preparation and if showing honey close attention must be paid to that particular show's schedule details. In fact, this applies to all the different types of honey you may produce.

The very first thing to do when you have a full, or nearly full, super of honey is to check the water content of the honey with a refractometer. It is a legal requirement for liquid honey in the UK to have a water content which is less than 20%. However, though legal, this is still quite high. May I suggest that a reading of somewhere between 17 - 18% is preferable if you are intending selling or showing the honey. Full instructions come with the refractometer explaining its use. This first basic check is essential to ensure that when your honey is extracted, filtered, stored in your buckets and jarred up, it will not ferment, and should retain its fresh clean taste for many months. Your honey should always be filtered before storage as this helps to stop natural fermentation occurring. If possible it should be stored at below 10 °C.

It is generally believed that when your honey is capped over by the bees, this ensures it is ripe and the water content is fine. This is not always so. Sometimes with a strong colony when a nectar flow is at its peak or starting to ebb the bees will rush the process of water extraction and cap full nectar cells over before they are actually ready, so always err on the side of caution and check it before extraction. If the bees have capped or partially capped it over and when you check it the water content is greater than 20% then you will have to extract it as if it had been capped in the usual manner, but then feed it back to them when the main flow has stopped to enable them to re-process it getting the water content down to the required level. Do this even if the bees are to keep the honey as winter stores, as a high water content can be a contributing factor to dysentery in bees.

If all is well, we then go on to extract the honey from the combs taking care to cut off the cappings that the bees put on the surface of the honey to keep it fresh. We can do this in a variety of ways but I use a reasonably sharp long bladed knife. With the frame of honey resting on a piece of wood placed across the top of a bucket allow the blade to cut just under the surface of the cappings and working in a downwards action cut all the way under the cappings which will then fall into your bucket. Repeat on the other side so you have exposed all the runny honey underneath. You are now ready to place the frame in your extractor. Repeat with other frames until the extractor is full then, slowly at first, start to turn the extractor and the honey will collect in the bottom of your extractor. The cut off cappings in the bucket can then be placed upside down on the top of the extractor with a muslin tied over the bucket top enabling the honey from the cappings to drain into your extractor. I place a plastic bucket under the extractor tap with either a conical filter hanging just under the tap, or a bucket filter placed on your bucket before we open the valve. This filter catches all the larger bits of wax and debris as your honey flows from the extractor valve into the bucket. If the filter gets blocked with wax and debris then close the valve, clean off the filter and re-start the process until the bucket is full or all the honey is extracted.

If you are extracting the honey on a very warm day in mid summer, the honey may be warm and free running. If this is the case it may be possible to pass it through the fine filter cloth straight away into a clean bucket. Before you do this skim off any debris on the surface of the honey with a large spoon. Always rinse your cloth under warm water first then shake off as much of the remaining water as possible before using. This enhances the ability of the warm honey to pass through the fine filter cloths. If the cloth is dry, it may clog on the inside and not run through smoothly. When it is a cooler and so more viscous, it may be necessary to gently warm the honey in the warming cabinet to enable you to pass the honey through the fine filter cloth.

If you have a thermostat on the warming cabinet set it at around 38/40 °C. Keep the honey in the warming cabinet until the required temperature is achieved throughout the honey. If you don't have a thermostatically-controlled cabinet but just a cabinet with light bulbs in the base this is also ok. Keep stirring the honey every hour and check the honey temperature with a jam thermometer or similar until the same temperature of the honey is achieved as above. Never use 60 watt bulbs as they are too hot and will burn or caramelise your honey. Use 40 watt bulbs in all cabinets, or slightly less. At 40 °C the honey should pass straight through the fine filter cloth from one bucket to another. Ensure the bucket is very clean first and again, before filtering, skim off any surface debris. It is always best to have another person to help with this procedure to avoid spills. Clean out the first bucket thoroughly and dry then repeat the process of filtration back into the first bucket. This honey then should now be very clean and can be poured into your settling tank very slowly. I often tilt the tank slightly on one side and pour the honey in down the side of the tank through the filter again. If this is done less air goes into your honey. Keep it in a warm place as you don't want the honey to get too cold. Leave it overnight to settle. This allows all the little air bubbles trapped within the honey to rise to the surface so when we bottle we get the least amount or air in our jars as possible. I have an old electric wine heater belt which I place around the tank base to keep the honey warm, but a nice warm room in the house should suffice.

We are now ready to jar our honey, but first read your honey show schedule to check which jars and lids to use. Most honey shows specify the use of 1 lb jars with gold screw lids but there are exceptions and so it is best to check. Always ensure that all the jars and lids are clean. Never assume that just because they are new they will be clean. I always put mine through the dish-washer first, and then ensure they are thoroughly dry. I then use a soft clean cotton cloth to give the jars a quick buff up prior to filing, inside and out before I run in my honey. Then we are ready. Tilt the jars as you fill them from the tank and run the honey in very slowly down one side of the jar, keeping any air bubbles to a bare minimum. I fill my jars ever so slightly over full just in case I have very small particles or a faint scum line that rise to the top. You can then skim it off the surface using the back of a dry teaspoon by rolling it gently over the surface, wiping and drying each time it is used. When the jars are filled set them in a warm place ideally on a window sill in full sun, or slightly shaded on very hot days. They need to stay there for a good week before the show, possibly two, and every day give the jars a sharp twist quarter of a turn around. This movement along with the warmth lets any remaining air bubbles rise to the surface and disperse. Any that persist around the jar neck can be lifted with the help of a thin straw or spatula.

We now need to use an LED torch to check through the honey. Remove the jars from the window sill, and very carefully keeping them upright place them on a flat surface. Then take your torch and hold it behind the jar and position it so you can look straight through the jar and the honey. This strong light will highlight any small piece of debris there may still be in the honey. These can be removed using a thin cocktail straw. Place your thumb over one end and lower the straw to just above the piece you wish to remove. When the straw is directly above the said intruder release your thumb causing a vacuum and this will suck up the piece into the straw. Place your thumb back on and withdraw the straw with piece inside. When all is clear and free from air bubbles with no granulation, with just very clear honey, your main task is done. Before replacing your clean lid just put a very small amount of Vaseline on the threads. This ensures the lid will not stick and it can be removed by the honey judge with ease later on the show bench.

There are three classes: light, medium or dark. It is very important that you check which class your honey falls into. If it is placed in the wrong class your entry will be disqualified. You need to either show your honey to a person experienced in showing honey who will be able to tell you what class it should be in. Or if your association has use of some honey grading glasses use those, but you will still need a person who knows how to use them correctly to assess your honey before advancing to the next step.

Sets of 6 samples of light, medium and dark honey. (D Shannon)

You must then refer, once again, to your schedule and check where your class label is to be placed. Some schedules say an inch above the base but the distance may vary from one show to the another. To get this measurement correct on each jar I have made a template. Just cut a strip of thin card cut to the correct size as the schedule states, let's say it is 1 inch. Make it long enough to go all around the jars you are using. Cut it to length and tape it together forming a ring. On this ring I place three marks. I place a jar into the ring making sure you can do this with ease and the ring is not too tight. Then I put a mark on the ring in black pen where the two mould seams of the jar are. Now take out the jar and lay the ring flat. Measure the position of the centre between the two seam marks and put a mark in red. You can now line up one of the black jar seam marks of any similar jar within the ring and the red centre line is always where you put the centre of your class label, making sure they are always in the same place and the correct distance from the base of your jar according to the schedule.

The main thing to remember with the preparation of your honey for the show bench is to take your time with the process and give yourself plenty of time to enable you to get the best results with your honey.

CHAPTER 3
Two Methods of Preparing Creamed Honey

Creamed honey, sometimes referred to as whipped honey, is delicious when processed correctly. It is a soft spreadable honey that doesn't drip or run, that can be spread on toast or crumpets etc. or can be added to beverages to sweeten in place of sugar. It is also good to give to children of all ages over 2 years of age as it will not run. Honey is not suitable for children under two in any form I would say, as the small amounts of pollen held within the honey can be harmful to them and in some cases cause an allergic reaction. For some unknown reason it doesn't seem to be sold and used as much as it should, but let us hope that we can change that.

All, or should I say most, honeys set or solidify due to crystallisation over time when stored correctly. There are exceptions but we are not concerned with those here. The honey should be clean prior to storage, sealed in a container that is air tight and full, with no air gap at the top and stored at the correct temperature to allow crystallisation to take place. This occurs at a temperature of 14 / 15 °C. The back of a cool garage or a cellar is ideal.

There are a couple of methods used to achieve a good creamed honey which I will attempt to explain. As with all honey, as explained in the liquid honey article, the first essential process is fine filtering to remove debris from the honey. This should be done when the honey is in a warm liquid form -immediately after extraction and before it is put into buckets and stored away. Any small dark specks or debris within the finished creamed honey will show through and make the jarred honey look very unattractive to say the least.

There are several types of honey that are very good for this process. Clover honey is excellent. Borage and Balsam honey can also be very good. But oilseed rape honey is the most commonly used now by most beekeepers when they do not have access to the other varieties. In most cases these honeys have to have the right balance of fructose and sucrose to aid the crystallisation process. A good and legal water content is essential as with all honeys but even more so with honey you wish to set smoothly with a very fine grain. High water content often gives course crystallisation. When the honey sets as a creamed honey this is not what is needed. You must have this smooth fine grained honey to start either of the two processes that I am about to describe.

1. Seeding honey for creaming

I believe this is the more common of the processes adopted to achieve a good creamed or whipped honey for beekeepers who have enough honey to fill a 28/30 lb. bucket. For this method you will require a bucketful of good quality liquid honey with a good aroma and taste. Warm it gently to 40°C. in your warming cabinet. If you do not have a thermostatically-controlled cabinet, then check the temperature with a jam thermometer at regular intervals. I still like to check the actual honey anyway to ensure it is correct. You will need around 3/4 pounds of a good fine-grained smooth creamed honey either purchased or swapped from another beekeeper to actually seed this honey. However before attempting to seed the liquid honey let the liquid honey cool to 22/23°C before attempting to seed. Alternatively you can use the same quantity of solid set honey from another bucket processed earlier and which has set to a hard fine smooth grain. Gently warm this three to four pounds of this hard set honey so it is no longer solid and achieves a smooth paste-like texture before adding to the liquid honey. Stir in the smooth creamed honey into the still-warmed liquid honey ensuring it is thoroughly mixed. Use a long-handled wooden or plastic spoon for this. This will enable you to get right to the bottom of the bucket ensuring a good even mix throughout. When this is completed allow the mixed honey to stand for about 8 hours and very gently stir every two hours if possible.

Next skim off any fine white scum or air bubbles that have come to the surface with your spoon. This honey is now ready to be either poured or ladled into your jars. Do not add the lids straight away when filled, but wait another couple of hours then gently stir the honey surface to remove any air bubbles that may have risen before adding your lids. These jars can then be stored away at a temperature of around 14/16°C - until the whole jar has set and it can be turned upside down without running at all. This normally takes around three days. The surface should look firm and dry, with the honey itself having the actual consistency of soft butter. Then label ready for sale. If this method is scaled up or down, 5 - 10% of seed is added to a liquid amount to achieve the same aims.

2. Using a bucket of set honey

This method of creaming honey is easier and no seeding is involved, however you require a bucket of honey that has been processed earlier and allowed to set. When once set only use the honey that sets to a very fine smooth and hard solid consistency. Place your set bucket of honey into your warming cabinet, and set the thermostat to around 35°C. Leave for a couple of hours and then check your bucket. You may have to do this several times. The consistency of the honey within the bucket you are wanting to achieve is approximately two thirds of the honey at the bottom wants to have gone back to liquid with the top one third remaining in a soft creamy but not a liquid state. When I am checking I squeeze the sides of the bucket and you can feel where the honey has melted and where it remains in a set state. When the bottom half is liquid cut into the honey surface with a long bladed round ended knife and cut to into quarters or eighths right through to the liquid. Then push one or two of the wedges you have cut down into the warmer honey and allow the liquid to move to the surface. This softens the rest of the hard honey quicker, allowing you to mix the two together with ease.

When you have achieved this two thirds to one third mix in your bucket then remove the bucket from the cabinet and place it on the floor on a plastic sheet in case of spillage. Now take your honey creamer, which you can purchase from any beekeeping equipment dealer. I prefer the hand operated one as opposed to the ones you place in an electric drill. I find the ones that attach to drills whip in far too much air into your mixture. As a result it takes longer to clear the bubbles and it has a tendency to form a scum on surface of the finished product due to the speed it operates at. Slowly, now push the soft honey on the surface of your bucket down into the liquid keeping the head of your creamer under the surface of the mixture at all times preventing air being taken down into the mix. Proceed to mix thoroughly till all the soft honey and liquid are evenly mixed together to an even consistency. Place your bucket in a warm place when completed and put the bucket back in the warming cabinet until the next day. Then I skim off any surface scum and very gently stir before bottling it. As with previous method leave the lids off and then ensure that the surface is free of bubbles by stirring the surface. Then jar and store as in method one.

Let us hope by having a go we can revive the popularity of creamed honey to more of our customers and friends.

CHAPTER 4
Cut Comb to Hive to Container

Cut comb exhibits in different packaging. (D Shannon)

Cut comb honey is, I believe, the crème de la crème of hive products. Producing a good quality cut comb is a way to show off your skills as a beekeeper in producing a fine product for sale to the public or for the show bench. A good reputation will be gained for a product that commands top prices from consumers. I will attempt here to explain the methods used to achieve this, including some of the pitfalls that may be encountered along the way.

When most people think of cut comb honey their mind goes directly to 'heather' honey; ling heather to be precise (Calluna vulgaris). This is distinct from the two other varieties of heather that grow within the UK, bell heather Erica cinerea, and Tetrelix, the latter a very low-growing small heather hard to find among the rest. Ling heather, however, is the most popular by far. Heather honey has a higher water content than other honeys and this should be tested before preparation for legal tolerances. Ideally this should measure no lower than 20% on your refractometer to a maximum high of 24%, but I prefer it to be below around 22%. Honey with a water content of above 24% is illegal to sell; it cannot be stored correctly and will ferment. It is far better to immediately feed this back to the bees for them to reprocess. Other honeys such as pure wildflower honey, borage, and Himalayan Balsam which are slow to crystallise, are also ideal for cut comb. You need to know the honey flows in your area, and then the principles below apply in exactly the same ways.

The essentials

The very first thing to remember when attempting to draw, fill and cap cut comb is the fact that this cannot be attempted without a brood box absolutely bursting with good healthy young bees, lava and eggs so that they are ready to maximise the nectar flow. The other essential is a vibrant new, preferably marked, queen of that season that is in full lay.

The whole process of preparation usually starts towards the end of June, beginning of July, when your hives are producing good strong queen cells. At this stage begin by selecting the hives and traits in the bees you desire for the job (good gatherers and bees that produce good clean cell caps). From these colonies you can also make up good strong nucs. from selected queens and produce good strong colonies to take to the moors at the beginning of August, leaving your other older breeding queens and hives behind.

Feeding these nucleus hives regularly is essential to give them the best possible start over the next few weeks; the newly available six-frame poly-nucs. with built-in feeders makes this so much easier. If you do not have these then a normal contact feeder with suffice. Fill your feeder with a light sugar syrup, then select the queen or capped queen cell and transfer into the nucleus, adding extra frames of both brood and bees as necessary. If the frames are light of bees they can be shaken from another colony into the box to ensure you end up with roughly 4 frames of bees. Always ensure they are free of varroa and other diseases prior to this procedure. Next transfer the nucleus colony into a full brood box when it is full and all space and frames are full of brood and bees. When you do this and with it being later on in the year, once again add a feeder to the hive and keep it topped up until all new combs that you place either side of the occupied combs are fully drawn out -that is if you are using foundation frames to fill out your brood box and not combs that have already been drawn. This will help ensure your new colony will be strong and full when the time arrives to proceed to the moors.

When supering for the heather several methods can be used and all are quite efficient. However, this is my preferred method. I always use new super thin comb foundation. This is a different quality of foundation supplied specifically for cut comb honey. I place this in my hives in the early spring when the oil seed rape is in full flow. I do not nail the foundation into the woodwork by the usual method, but prefer to use hot molten wax to stick the foundation into place on the top bar of the woodwork after first sliding it into place down the side groves in the frames. This ensures a better hold when dry, as the very thin sheets can sometimes slip away from the nails if tradition methods are used. It does not drop when the weather gets hotter or fall out of place or distort. When these frames are full, however, great care is taken when uncapping them and especially when extracting them; this must be done slowly to avoid any damage. Once extracted, I take the frames back to the bees to clean out before storage for the heather crop in August. This method avoids the bees having to work twice as hard to draw out the comb on cool days on the moors, and means they can concentrate their efforts on filling and capping the lovely new frames ready for cutting. If in your area you have another good source other than oil seed rape then this method can be applied to any such crop where the bees can draw out good combs.

If I do not get enough drawn comb by this method I cut foundation sheets in half length ways, this method is called the starter bar method whereby when they are fixed in pace the bees draw them out from the top and then hopefully continue to draw them all the way down to the bottom bars into full frames when capped. I fix them in place using the same method as before, with hot wax applied to the top bars then the bees will hopefully draw them down and fill them with fine Heather honey. Only one super box is placed upon each hive when I take the bees to the moor. The exception would be when a colony that has exceeded itself and is totally bursting at the seams. I then add another super to provide space for the bees and extra brood when it hatches. This would be placed above the first super as a second super. When I add supers I always keep my queen excluder in place on top of the brood box. I do not want the queen coming up into my supers to lay, as this detracts from the quality and hygiene of the finished comb.

The hives are taken to the moor during the first week of August but exceptions have to be made if there is warm or wet weather throughout June/July as the heather could flower earlier. (It is important to keep checking on its progress).

On the moors

Ensure your hives are very full of young flying bees and brood at all stages, with at least one super on each. Then use two straps in a criss-cross to prevent any twisting in transport and to create stability while on the moor. It is always best to have prepared in advance a good location which is well drained and in a sheltered area. Getting a vehicle bogged down on the moors in a remote area is very easy, so be aware.

Never place hives on top of a hill or mound, as exposure to the cold winds and rain is very detrimental to your bees and that all-important honey crop. Always place your hives at the bottom of the moor so the bees are going up hill when going out and down hill when coming back, saving them energy. The hives generally stay on site for about two to three weeks depending on the conditions and once the bees have completed their task of filling your supers they can be brought home.

Treatment of the bees against varroa and nosema and topping up their food resources is vital to sustain the colony after you have taken off the crop. Fumidil B is no longer available for use against nosema but other new hive cleansers are readily available to mix into your winter syrups when feeding and are very effective. I take this precaution due to the very high water content of the heather honey and the knowledge that the bees will have stored some of this in the brood box; it often does not store as well through the winter months causing some cases of dysentery.

Back in the apiary

On return the task of preparing that quality product begins. At this stage you should have purchased a couple of packs of cut comb containers and their decorative sleeves from your local stockists in readiness for filling. The first thing to do is to ensure you have a good clean working surface to prepare your cut comb. Clean and disinfect the whole area with a food quality antibacterial spray, including any knives or other utensils to be used in the operation and then dry thoroughly. Place your frame of sealed heather on a flat dry surface and carefully cut around the inside of the frame releasing the comb from the woodwork. This is when 'gluing' with wax is preferable to nailing into the woodwork. For cutting the comb, I use a simple template made from a small piece of plastic with a handle attached at one side. This should be the same internal dimensions and shape as your cut comb container, so that it represents the piece of comb to be cut. When this method is used it allows greater consistency and a nice clean cut piece of comb every time. Only use the area of comb that is completely sealed.

Place your template on the surface of the cut comb at one corner then mark around its exterior with a hot knife; this needs to be regularly cleaned it in very hot water to ensure a clean cut. I prefer to use a cheese wire when cutting comb as this is very thin and allows a better edge to the sides. I also have a large mesh cake stand to hand in front of me, on which I place the cut combs while the surplus honey from the newly cut edges drains into a drip tray. This ensures a nice clean finish when they go into the containers, with no honey running in the bottoms of the containers. If you are planning to show your cut comb, it is important to cut all the capping in one direction.

If the frames have been drawn out to their full extent, then each shape you cut out should produce a piece of cut comb that weighs between eight and eight and a half ounces. Before placing them into the containers and sleeves, check the weight of each one. If selling to the public, write the weight on the cut comb container and your contact details. The comb surfaces should be clean and dry and your containers should be free from 'runny' honey. If they are going to a show then the weight is crucial: 8 oz each before they are placed into the container with a nice even surface. Once completed, place them into a large plastic bag, in a strong cardboard box and pop those into the deep freezer where they will keep very well until required. Then bring them out two to three days before they are to be used and they should be perfect. Or simply sell them directly to the awaiting public.

I leave the labelling until the comb comes out of the freezer and the containers have dried as the damp created during thawing causes the sticky labels to peel off as they dry out. A secondary, and excellent, product from cut comb honey other than the traditional cut comb is chunk honey, another very popular product. This is when a chunk, or piece, of cut comb is carefully cut out of the full comb as before, but the dimensions of this piece of comb should be cut to the same size as to just fit inside a standard 1lb. honey jar neck, leaving virtually no or very little gap where the corners of the comb square touch the inside of the jar neck as its lowered into place. A template can be made ensuring it fits the internal dimensions of the jars neck before cutting takes place. It should also be long enough to fill the jar from top to bottom when in place with no gap between top and bottom of the jar, so check this measurement also. If it's too short it will float upwards when liquid honey is gently poured in around the chunk to fill the jar completely to complete the process. I always think this looks best when a chunk of heather honey is used and the jar filled with a very light-coloured honey as this produces a contrast to the reddish colour of the heather honey chunk. However, this is not essential and other types of chunk honey can be used but not quite with the same effect.

Important tip

Heather honey cannot (**normally**) be extracted from the comb in the usual centrifugal extractor because of its very unusual gel-like properties (thixotropic) that hold it into the cells more firmly.

However, several honey looseness - either roller type ones or blocks of wood, each with rows of spikes, can be used to both break the cappings and agitate the thixotropic honey to enable it to run out of the cells when the frames are put in a honey extractor.

Heather honey loosener. (J. Phipps)

CHAPTER 5
Sections

Section honey - the best of all products from the hive. (D Shannon)

Square sections

Section honey has for many years been considered the absolute best product from the hives, and the premium in cut comb production. The person we have to thank for first producing this product was a very eminent Russian beekeeper called Peter Prokopovitch over 170 years ago. Sections both round and square are the purest form of honey to come from the bee hives as they are virtually untouched by human hands from when we put them into our hives until they are consumed by the purchaser. Square sections were the first type to be produced and then, more recently, the plastic round version made section production even more popular with beekeepers in the 70's as they are so much easier to produce, but still remaining a premium product.

Components required: section box, section square wooden strips, super thin wax squares for sections, Plastic dividers, tension board and springs x 2.

If you go down the path of doing it yourself a section box is required, as square sections do not fit into a normal National super. They require a special box designed especially for the purpose. The National section box holds 40 sections when full. This box is open at the top to allow the individual sections, when once assembled, to be fitted in. It is slatted at the bottom to allow you to stand the sections on them whilst assembly is taking place, plus the spaces between that allow the bees access to the sections from the brood box below when they are finally in place in the hive. They come in a bundles of 10s I believe, but 40 are required to totally fill the box. I would also allow a few extra for breakages whilst assembly is taking place as they are quite fragile. The section outer cases come in flat strips, made from fine light good quality timbers and are notched out to allow them to be folded into the square shape around the wax square. They are also slotted on the inside to hold the wax in place. However, do not try to fold them while the wooden strips are dry! They will snap. First they require soaking in a bath of warm water, say 10 at a time for 10 minutes. This makes them soft and pliable. Take a wooden section strip and fold it with the notched out lines and slots on the inside forming a square. One of the four sides is split into two halves and this is the top. One edge is notched out on both these pieces and slots into the opposing corner locking the square in place. Only fasten one strip in place, then slide your wax square down into the pre-cut grooves on the inside of the square, sometimes this may have to be trimmed slightly for it to fit snuggly. Take care to make sure that you keep the wax embedded in the slots, then close the last hinged strip to form a completed square with the wax inside. Repeat this until you have a row of 4 sections in the box then place a divider, separating that row from the next. If this is not done the bees will form brace comb between the two and stick everything together in a total mess. Repeat this process until the box is full. You will then have a small gap left over. This is when we use our tension board but don't forget to put in a last plastic divider first separating the two. After the board use the two section box springs to slot in between the tension board and the box wall, this secures everything and keeps it all tight and secure.

When this is completed, the honey flow is on and your hive is bursting with bees. You are ready to go off to the moor, praying for a fair wind and good conditions to allow your bees to fill your lovely sections up. Each full section will weigh between 14/17 oz. when full. When the flow has stopped, bring them home, take off your sections and clean the outside of the wooden casings thoroughly. Weigh them, place them into presentation cases, add each section's weight to the carton as well as your address, then they are ready for sale. If you are not selling them straight away, they can be stored in a freezer. Wrap in cling film in batches of 6/8, then seal them in a box before placing in freezer. Here they will stay in good condition for months. Always allow a few days to defrost before selling on. With these products only use and sell the fully-sealed sections. Unsealed sections should be pressed out for jarring. Alternatively feed them back to the bees. If the bees clean them out well, use them next year to save the bees drawing out the comb again.

Round sections

As with the wooden sections, these can be bought complete in the box from your beekeeping stockists. Once again they are not cheap but potentially very profitable. When full and sold on they achieve a premium price due to their excellent quality. The big difference is that the rounds cases are all plastic, not wooden, and very easy to assemble. You can also clean them up again for future use, only having to purchase more rings and labels, so there is less outlay the second time around. You will however still need a section box.

ROSS ROUNDS™ equipment consists of brown moulded plastic frames into which, when split in two, you place your white plastic rings. There is one in each circular section, a total of eight per frame. A good

Round sections are gradually becoming more popular. (D Shannon)

tip is to smear very lightly the outer rim of your rings with Vaseline to stop them sticking prior to putting them into the frame. Then place a thin sheet of wax foundation on one side of the frame covering over the top of 4 of the rings. Then carefully fold the two halves of the brown frame together to form one unit and click together. These are then placed in a wooden super or rack. When full, add the tension board and springs and they are ready to be filled with honey by the bees. Plastic covers are available made to fit the top and bottom of the sections when full, as well as a wrap-around label which completes the package. Covers can be clear both top or bottom, or a clear cover on top and an opaque cover on the bottom. This last arrangement is preferred by some commercial beekeepers. All can be purchased individually including the wax sheets that are cut to fit the four round sections in one piece so you make up a row at a time and not individualy. You will also need a pack of labels. It is often said it is easier to produce rounds rather than square sections as the bees fill them better with being round. Sometimes with the squares the bees don't fill the corners correctly, which is not a problem with the rounds.

When the rounds are full and the flow has finished, take off the box and take out a section of 4 rounds. Split the dark brown section cover into two halves carefully with a blunt knife and go around the clear plastic section ring to enable you to remove it and its contents. You may need a very thin knife to slide around the outer of each ring to penetrate and cut cleanly through the wax in the middle. Clean up the outside of the ring, weigh each one as you go and place it into your clear plastic show case base and put on the lid. Each round section, when fully sealed, should weigh between 12 and14 oz. You can buy the labels in a roll of as many as you require. Peel off the protective cover and stick it to the outer rim of the clear show case. Add your rounds weight to the label and, if for sale, your address details. This acts both as a ordinary and a tamper-proof label as well as ensuring freshness of the product within. If you don't wish to sell the sections straight away place them into a stack of 10, cover them with cling film and then into a box in the freezer. They will stay fresh like this for months. As with the wooden sections just remove them 24 hours. before you need to consume, show, or sell them.

I do hope you have a go at this method of honey production as sections are great product, great on the show bench and very popular with the public as a tasty treat of the highest quality.

CHAPTER 6
Cakes and Confectionary

Mix all ingredients thoroughly to a smooth consistency. (J Phipps)

The plain honey cake

As with every section in a honey show, cakes are not an exception, so it is important first of all to consult the bible, i.e. the show schedule, for detailed requirements for the class.

Read through the section on cakes carefully so that you digest and follow all the relevant instructions to the letter. This is vitally important with all the classes, none so more than with the cakes. Make sure before you start you have all the right ingredients to hand and, very important, check the stipulated tin size that the cake has to be baked in. If it states a 7 inch tin then it's no good using an 8 inch tin and thinking "oh it will be ok". NO it won't. All judges will read the schedule and the first thing they do in these classes before even examining the cakes is to check the cakes size with the size stipulated within the schedule. If it's the wrong size it will be discarded, classed as not to schedule. So, correct tin size. If its not a new tin ensure it is perfectly clean before advancing, then line your tin with greaseproof paper and lightly grease it before placing it in the tin, the base also. Most schedules say use a greased tin, but this is the best method as it helps to protect your cake from burning when cooking. Once all this is prepared, set your oven and turn it to the recommended temperature.

Now for the ingredients. Once again consult the schedule for this and the method laid down to accompany it. It is very important to measure every ingredient correctly. Only use fresh new-laid eggs. Pass the measured amount of flour through a fine sieve Photo 6 before adding the other ingredients and always use self-raising flour, not plain. I even add a good pinch of baking powder also at this stage to help with the rise. Then follow the instructions in the cake making method in the schedule. Make sure you mix all the ingredients thoroughly to a very smooth consistency before placing it into your tin. I also never use warm margarine or butter but keep it at room temperature or slightly below.

It is always best to have a dummy run with cakes to get the right setting to your particular oven with regards to baking times and temperature. as all ovens vary slightly and a slightly lower temperature is better than one that is too high.

Keep an eye on the cooking time especially when it is almost cooked. I would suggest that after an hour at the cooking temperature the cake is tested with a skewer. If it pulls out of the cake clean then it may be done. On the other hand if its sticky on the skewer then it requires more time. It should be a lovely golden colour all around and very even across the top. Allow it to cool slightly then remove from the tin and stand it on a cake tray until completely cold. It can then be wrapped in tin foil and either taken to the show or if done in advance wrapped, placed in a cake tin and stored in the freezer until the day before the show. Defrost the cake gently for 24 hours before it goes to the show. Always remove any lining paper from the cake prior to staging and please also ensure you have the cake properly staged. If it states "on a doily on a paper plate" then ensure it is. If it states one label to be placed on the plate, ensure it is. Also if it states to be placed into a clear plastic bag with a label on the bag also then please ensure this also. This cake is now ready to show.

Honey cakes ready for judging. (D Shannon)

Fruit cakes

For a fruit cake all the above information is to be carried out for this cake as for the plain cake. The only variable is the fruit that is to be added. Pay particular attention when and if it states the type of fruit to be used and always adhere to this. Follow the method when mixing all the ingredients together. One tip with your fruit: when you have weighed out and sieved your flour, a little should be taken from it and sprinkled over the measured amount of fruit coating all the fruit with a fine dusting of flour before adding it into the mixture. Once again, always ensure that the ingredients are thoroughly mixed before adding the fruit. Then fold the fruit into the mixture making sure its all well-mixed before pouring into your tin, prepared in the same way as above. Some people pre-soak their fruit in various liquids to plump up the fruit and make them tastier. I would not recommend this as it adds weight to the already weighed out fruit and can make your cake too heavy.

With nearly all cakes in shows this section is the only one that very often states that the honey used does not have to be the makers own produce. This means shop bought honey may be used. A mixture or blend of honeys are often best for both. I personally prefer a wild flower honey for my plain honey cake, but a mixture of wild flower and heather for my richer fruit cake. Another thing about fruit cake is that, with them having a greater amount of contents, the cooking process takes longer to ensure good results. So a slightly lower temperature then advised and a longer slow cooking time will be required. Take care not to burn the outside or top of the fruit cake. As with both cakes a trial run will help regards getting the temperature and cooking times correct to perfection. I do hope this prompts more contestants to come along and participate in our honey shows, especially as these are both two classes that often don't require the honey to have been produced by the entrants own bees, hence the competition is open to everybody and not just beekeepers!

Honey confectionary

Honey confectionary/honey sweets as this class is often termed is another section where anyone can enter as in most cases it is not restricted to beekeepers alone, unless stated in the schedule that the honey has to be produced by the exhibitor's own bees. Now this class varies from show to show so once again reading thoroughly the actual show schedule is vital prior to exhibiting. I personally don't think that any flour-based receipts should be used in this class eg. scones, shortbread or such like. It is a sweets/confectionary class. Chocolates, fudges or a honey marzipan based recipe should be stipulated. However, if the schedule is followed all will be well. Whichever outline ingredient you go for in this class, the main thing is that the end product actually tastes of honey and not everything else. This class gives the exhibitor great scope to produce nice uniform products of their own choosing. Most schedules give an overall max/min weight to be exhibited and also most depict the number of entries to be produced from the same recipe for example, 3 to 6 pieces per plate. Just ensure all your sweets or pieces within the exhibit are uniform, look inviting and taste of honey. The rest in this class is up to you.

I hope this encourages you to come along and enter our honey shows in these particular sections even if you're not an avid beekeeper. Who knows you may get to win a top prize and rosette for your troubles.

CHAPTER 7
Meads, Dry and Sweet; & Fruit Melomels

Let us begin this section by stating that there are many variant recipes for making mead. The text that follows is a helping guide and describes my personal way of making a good mead. And now, having said that, let us get started by listing the equipment you will need to make mead successfully.

You will require:
- a glass demijohn, that holds approximately 1 gallon.
- a wine hydrometer to check the progress of fermentation.
- an air-lock.
- a rubber or cork bung with a hole in it through which to insert the air lock.
- a solid rubber or cork bung.
- a metre of wine syphoning tubing.
- a plastic funnel.

All of these can be purchased at your local brew shop, such as Boots or the Wilco's stores and they are not overly expensive as they can be used time and again. When all these are in place, next comes the ingredients to make your mead wine. These vary dependant on the type of mead you are wishing to make, but they are basically water, honey, yeast, a yeast nutrient and camden tablets.

Let us begin with making a gallon of a dry mead:
You will require between 2 to 3 lbs of honey, a nice floral honey is good for this and you don't have to have your own hives of bees to source this - a shop bought floral honey is quite adequate. Ask your brew shopkeeper for a dry wine yeast and then you are ready to begin. De-chlorinated water such as clean spring water or bottled still spring

water is best to use with all meads, not tap water. Pour a good cup full of this water into a jug and add to this half a cup of honey and stir till dissolved. Now warm it to 68/70°C using a jam thermometer to check the temperature. If your purchased yeast comes in a packet use just under half of this and sprinkle it onto the surface of your water/honey mix in the jug. Cover with a clean cloth and leave it for around half an hour for the yeast to activate. You should have a bubbly frothy top formed on the liquid when ready. Now pour the remaining honey into your demijohn, add a teaspoon full of yeast nutrient and add your water already warmed to 68/70°C. At this stage I like to add just a squeeze of lemon juice to the mix.

Don't fill the demijohn to the full, just to the three-quarters mark. Then add the yeast mixture to the brew. Give the mixture a good shake or stir to ensure that all of the honey and contents are well mixed together. Top up now to about two inches from the top. Now fill your air lock with about an inch of water, push it well into your bung with the hole and place this into the neck of your demijohn. The demijohn should now be kept at around the start temperature in a warm place to ensure the yeast keeps working thus turning all the sugars of your honey into alcohol. A good thing to buy is a wine mat or belt. These are both types of electric heaters and can be bought from your brew shop and they ensure your brew is kept at the right temperature. Stir the liquid twice a day for the first three days then leave it to ferment in peace. This process normally takes about two to three weeks. Look at your air lock after this period and there should be no more bubbles plopping in the airlock. If there are, leave until it has stopped. Now take your hydrometer and tie a short length of cotton to the top of the narrow measuring tube. This allows you to drop the hydrometer into the demijohn after removal of the air lock. For a dry mead the gauge should give a reading of between 0.990 to 1.005/6. Great care should be taken when doing this that you do not move the demijohn in any way, as you don't want to disturb the layer of yeast that has formed on the bottom of your brew.

If the required reading has been achieved it is ready to "rack off". This means placing your demijohn onto an overturned biscuit tin or something to elevate it higher than a large bowl into which you are going to syphon off the liquor from the demijohn. When you get near the bottom of your liquid tilt the demijohn carefully, but stop when any of the yeast mixture comes through. We need the liquor to be as clear as possible. This done, clean out the demijohn thoroughly with a bottle brush and hot water. When it's clean, place your funnel into the neck. I place a spoon down the side of the funnel into the top as well. This allows the air to come out when you start to pour the mead back into the demijohn. Pour it slowly. I then add to the mixture two crushed camden tablets, shake the mixture well and replace the air lock for a couple of days until all further fermentation has stopped. At this stage I place the demijohn on a wedge under one edge thus tilting it over to one side. When this is done I rack it off again. By placing the wedge under an edge you should have a clear side at the bottom of the demijohn from which to siphon, so keeping your liquor clear. After this replace the airlock with your solid bung and store in a cool place. I put mine on the floor at the back of the garage. If your intentions are to drink it, it will be ready in about three months. However, if you are going to show it, it needs to be left for about three years. I rack mine off every 6/8 months as mead will continue to drop sediment throughout its lifetime. If everything has gone to plan you should have a nice dry mead with a flavour not unlike that of a dry wine. However, you may not get the same aroma as with a dry wine, as most floral honeys have a very delicate flavour that is sometimes lost in a dry mead. It should have a good alcohol content of around 7-12%, be pleasant on the nose and pallet and not have a sulphur taste or be too harsh.

Sweet Mead

With a sweet mead the actual ingredients are the same but the quantities are different. I will go through all possible changes but the process remains largely the same as with the dry mead.

When making a sweet mead there two main changes to note are:

1) the amount of honey,
2) the type of yeast, and often the addition of tannic acid.

Mead - sweet or dry, melomel, beers - there are several classes for those who wish to show beverages. (D Shannon)

To make a good sweet mead we need a stronger flavoured honey and the best of these I feel is heather honey. However, this can be blended to give a more rounded enriched finished product with other honeys. Balsam and clover also provide a good flavour. We need to use around 4/5 lbs of honey per gallon to ensure a good fermentation. If blending, I use 4/1 heather to balsam mix as the balsam has a citrusy after-flavour to contrast with the rich heather taste. You will also require a sweet wine yeast - a Sauterne or Bordeaux yeast is good for this. Follow the procedure as above, allowing the yeast to activate thoroughly before mixing in. Add your honey then your yeast nutrient and a squeeze of lemon or lime juice, (half an egg cup maximum). At this stage some people use a pint of cold tea to replace tannic acid and to add body, but this is a personal choice. Keep to the same temperature and maintain it throughout the fermentation process. This may take considerably longer with a sweet mead in some cases but not always. When the bubbles have stopped take your hydrometer reading as before, but this time it should read between 1.012 to 1.020 for a good sweet meed. If it is above this reading of 1.020 it is no longer a sweet meed but a dessert mead, that is the defining difference. Rack off and store as before and if showing it rack it off every 6 months and show after 3 years of age.

The result with a sweet mead is that you should have a very nice distinctive sweet honey aroma and the pallet should retain a full-bodied and rich intense complex taste. A good test of its "legs" (or alcohol content) is to pour some into a very clean warm wine glass, swirl it

. . . . making mead is a long and involved process including fermenting and several sessions of filtering and racking , before laying down for two or three years.
(D Shannon)

around the glass and you should see with both dry and sweet meads where it has clung to the sides of the glass in large waves as you have swirled it around and the mead has re-settled again. This is a lot more apparent with the sweet meads as the alcohol contents are much higher, double in some cases.

Fruit Melomels

A Melomel is a mead that is made with either seasonal fresh fruit, or from fresh fruit juices, sweetened only with honey. If you intend to use, as I do, seasonal fruit you are going to need a mash tub. For this I cut a hole in the lid of a 30 lb bucket the same size as to fit into it my airlock bung giving it a tight fit so no air comes either in or out only through the airlock. This done you are ready to begin. It is then a matter of selecting the type of fruit mead you wish to make. Most Melomels are either medium sweet or sweet due to the sugar content within the fruit itself plus the honey used. Any type of soft summer fruit can be used to make Melomel.

Lets use as a for instance blackberries. Take 3 lbs of blackberries and put them into your mash tub, after washing them thoroughly. Now take 3/4 lbs. of honey, floral or similar. I then use a potato masher and squash the fruit to get out all the juices and flavours. I then add a teaspoon of yeast nutrient a squeeze of lemon and top up this with around 6 pints of spring water warmed to 68/70°C. I have already activated my yeast by the same method as previously explained, and use the sweet mead yeast for these. Now add all together and stir with a wooden spoon. Do not use a metal spoon when mixing yeasts, wood or plastic is best. Place on your airlock and lid but take off the top from the air lock before putting on the lid. If you leave it on the pressure will pop it off and you will have to re-fill it with water. If at this stage I have a little excess, I place it in a wine bottle with a wad of cotton wool in the top. Then treat this as I do the main lot. Stir twice a day for 3-4 days then leave for a further week until this first vigorous fermentation has eased. Then strain off the old fruit through a muslin cloth into a bowl. It will still have yeasts within the liquid at this stage. Get your demijohn and funnel and gently pour your liquid into it. Don't forget the spoon as before.

This will then need to be kept like before at around 68/70°C. until the fermentation has stopped and bubbles ceased. It generally takes another two to three weeks but it can take longer. This then will require racking off as before into a bowl then back into a cleaned demijohn. If it does not fill the demijohn to an inch from the neck rack off the excess you put in the wine bottle and top it up. You may have to repeat this again in a months time when done and all fermentation ceased. I crush up a Camden tablet into it and leave with the solid bung it in the rear of the garage to mature with the rest. I do hope the explanation of mead making will encourage you to have a go.

CHAPTER 8
Observation and Nucleus Hives

More shows are making space available for observation hive. They are always of interest to visitors to the shows, especially non-beekeepers who are fascinated to see bees close-up. (J Phipps)

Not all shows now have either or both of these classes within their schedules. At the Yorkshire BKA annual autumn show, the Countryside Live show at the Harrogate show ground do not. However, at most larger major shows they are both included. The Great Yorkshire Show honey show run by the Yorkshire Agricultural Society have both of these in their annual July show. Although with the building or their new and splendid honey house officially opened at the 2014 show, the observation hive section had to be temporarily postponed until the actual placings of the staging within the new building to accommodate this class has been established. This was finally finished in 2015 so everything should be in place from 2016 onwards.

The Observation Hive

The preparation for the exhibiting of an observation hive is crucial and not something I feel to be taken on by a beginner beekeeper with little experience. There are several methods from which to select your frames of brood with a laying queen for this task, but I prefer to take my brood frames from a nucleus hive for this purpose. By doing it this way it is far less disruptive than depleting a full colony of its queen and some brood, especially when the show may go on for two or three days. Additionally, the hives have to be staged the day before the show adding another day to the whole thing.

When preparing your nucleus hive a new queen isn't essential for this task. This is because most observation hives only have two brood frames in which the queen can lay. New queens are virile and lay at a very fast rate, filling all remaining cells within possibly a day. This then leaves her with nowhere else to lay and she will stop laying. If an older queen is used from the year before and placed in a nucleus, her laying capacity will have become somewhat depleted - but as long as she is laying, even be it very slowly, this will be an advantage overall.

I start the process of providing material for an observation hive when I make a split after queen cells begin to be produced in a good selected hive that is free from any disease. It should be free from American Foul Brood (AFB), European Foul Brood (EFB), and VARROA. I take out the old queen and place her in the nuc box and allow a queen cell to mature and a new queen emerge in the original hive. I then place the nuc next to the original hive. I try to ensure that this nuc is more or less full by the time of the show. If its overfull a week before the show I will replace a few frames to keep the queen laying and ensure they are not overcrowded. I then select my frames to go into the observation hive as close to the show day as possible. Whilst looking through the frames, the frame with the queen actually on is a good start. I try to use an unmarked queen and mark her with the colour of that particular year when putting her into the observation hive. This frame, plus another one, are put into the observation hive. Do not allow the hive to be overfilled with bees - allow some of the flying bees to fly off by gently tapping the frames but keep your eye on the queen as you do this. It would be a shame for her to fly off at this stage and get lost. You are looking for frames with good food stores around the top of the comb in the form of sealed and unsealed honey, then a good visible arch of pollen, followed by brood at all stages on each frame. Additionally, a good clean frame full of honey stores is needed for the top of the hive.

When all these are in the hive you then have it to mark up. Read your schedule as this very often says what and how things must be marked. In most cases little sticky paper arrows are are stuck onto the outside of the glass pointing to specifics within the frames: honey, pollen, sealed brood, unsealed brood, lava and eggs.

This is then ready for the show and hopefully a prize.

Nucleus Hives

The preparation for a nucleus hive starts when you do most of your your splits in late May time. Select from a good gentle and good-laying parent stock. Ensure all the way through that they are both disease free from AFB, EFB, chalk or bald brood etc and pests particularly Varroa. You want very clean frames well drawn-out with no holes in the combs.

On selection you will require:
• a well-marked queen with the current year's colour,
• a very good brood pattern on at least 3.5 / 4 frames of bees,
• a frame of stores.
• All the frames should be matching in type. It is always best to "bleed off" • most of the flying bees prior to taking the nucleus to a show. This is done • at the time of the final sorting of the nucleus, when your ensuring all the • above are in place.

To do this move the main hive over on its stand to nearly cover where the nuc has been. Move the nuc to another area within the apiary and most of the flying bees will go back to the original place. They will eventually go into the hive you moved over. This leaves your nuc ready to take to the show. In some shows such as the Great Yorkshire Show, after the judging on the first days opening morning, the nucs are used as an education tool, alternately, to demonstrate to the public the inner workings of a beehive. There is normally a section within your entry form asking for you to sign if this is ok. Every hive will be rechecked again by the bees steward to ensure they are friendly and disease free. Nasty bees or diseased bees will be closed up, removed and WILL not BE judged!

This is an area in which we are desperately trying to get more entries and I will appreciate anyone's help in boosting numbers by entering.

CHAPTER 9
The Display Class

A majority of the Honey Shows, County Shows and larger events now have a display class. This varies from show to show and can consist of virtually anything that the organisers prefer to see in the show.

When entering the display class it is very important to thoroughly read the description of articles required for display within the schedule, and ensure that you stick to them. Do not add anything else unless it states that a decoration may be added. If it doesn't then don't! Also, it's very important to stick to the precise size of the display board area given within the schedule. If it states (example) 350 mm x 350 mm then ensure that your board is no larger or smaller. You may cover it to aesthetically enhance it, or buy one already prepared such as a silver or gold board. On your board prepare your display articles individually, one at a time, ensuring they are all presented to the very highest standards.

Some schedules state wax models, and some say 5 or 6 per display consisting of different models. The models can be made from decorative candle moulds, however, you leave out the wick to make a model instead of a candle. As with all wax products though they must be produced from very clean wax, and perfectly moulded.

Another, for instance, is 'Products of the Hive'. Once again displaying a set number of items upon a given area, but often leaving it up to the individual to decide on the actual products they wish to display. These can vary from a small pot of honey, beeswax polish, a decorative candle displayed in a fancy holder, a beeswax model, honey vinegar, or handcreams made from beeswax. All of these items are acceptable along

Display Classes allow beekeepers to show their diversity, however, exhibitors must not use more than the space specified.
(John Phipps, Nation Honey Show, London)

with any other hive product that you may wish to include, unless the items to be displayed are individually itemised within the schedule, in which case these must be adhered to. In some top shows they leave the display itself entirely up to the individual. This means that you can use your own creativity to produce a display of your choosing, either from set pieces as mentioned earlier, or a more creative and abstract art form from beeswax.

Whatever the display, correct display area size and well turned out individual items are essential to catch the judge's eye and achieve a top prize card in this class.

This is a great class for someone with flare and imagination, so go on and have a go and let's see more entries within this class in future.

Let's get creative!

CHAPTER 10
Beeswax

A beautiful and elegant display of beeswax candles. (D Shannon)

When selecting your beeswax for the show bench, be it for a block of show wax to be moulded to a perfect finish or for a commercial block of wax class (which in most schedules does not need to be moulded to perfection) or producing a pair of matching beeswax candles, you need to start with the best ingredients. By this I mean very clean, well-filtered wax melted at the correct temperature so as not to darken the colour of the wax and thus keep its natural clean beeswax aroma. How do we obtain this wax and then how do we go about cleaning it this very high standard?

The very best wax to use for this is 'cappings' wax - saved from when you have extracted your honey. Don't use an un-capping fork this process - use a sharp, warm, round-ended knife and cut just under the cappings surface taking them off in a thin sheet of wax and honey.

Allow all of the cappings to drop into a bucket until all of the honey frames are uncapped and the honey extracted. When this is done, place a fine muslin over the bucket top, tie it off with a string and tip it upside down over your open extractor allowing all of the excess honey to run off into your extractor leaving you with just the sticky cappings. These are then thoroughly washed, preferably in rain water from a water butt; as this is softer than tap water.

Repeat this process until all of the honey is removed from the cappings. Now place them into a pair of nylon tights and hang them on the line on a fine day to dry. Give them a good shake again to move the cappings and to help to remove any remaining liquid.

This is now the wax you will use to make your show wax. You can also add to this any brace comb or wild comb you have removed and saved over the season, however, only use the very clean wax and discard any dark or dirty wax as this will taint the rest.

You are now going to need a large old pan. Fill half full with clean rain water and bring to a rolling simmer. Next, begin to add your wax stirring as you go. When it's melted allow it to continue in a rolling simmer, this allows all the small bits and any other detritus to fall below into the water leaving the clean beeswax on the top. Personally, I cover a couple of tea trays with greaseproof paper, take a ladle - preferably one with a small pouring spout - and carefully skim off the clean melted beeswax from the surface. As I do this, I pour it into the trays I have prepared and take care to not pick up any little bits. When the trays bases are all covered stop and allow the wax to cool. Then peel it off the greaseproof paper, you can see easily any little black specks and pick them out leaving only the very clean wax.

It is with this wax that we will make show cakes, commercial wax blocks or candles. For all of these, melt the wax once again but this time in a Ban-Marie. You will at this stage need to read your show schedule and weigh out the clean wax precisely to the weight of your show or commercial block. Don't turn on the heat to the pan yet. Prepare your mould - a glass pyrex dish is usually used for this. If a specific

Great care must be taken when melting wax. Always use a double boiler and remember wax is highly inflammable. (D Shannon)

thickness of wax is needed make sure your mould is big enough to achieve that thickness when the wax is poured. If it's too big reduce its size. Put a very small drop of washing up liquid on your finger and rub it all over the inside of your dish covering all the surface. Then place it upside down in a warm oven! You will also need a piece of glass cut big enough to cover your mould top and put them both in the oven

61

together at around 60°C. Now cut a piece of surgical lint and place it on top of your inner pan covering one half of the pan top and secure this in place with a strong elastic band. Put all the required wax into your pan and turn on the heat but not up full, allow the water to come to a slow boil melting all the wax. Now take the mould out of the oven and allow it to cool a minute, then pour your wax into it through the lint. When the pan is all empty and the mould full, leave it for a couple of minutes then take out your piece of glass from the oven and place it on the top. Next cover this with a heavy towel and leave it until it's cool. Now take off the towel and glass. Fill the mould top with water and place it in the fridge. Leave it there to cool completely, the wax should then pop out of the mould.

This can then be polished with another lint piece under cold water - a little soap may be used. You should now have a good show wax piece.

This same process can be used for commercial wax but the moulding need not be as good.

The exact same procedure is used for the candles but you pour the wax into candle moulds instead of a dish. Don't forget to dip your wick in the molten wax before you put it into your candle and use the correct wick size for your candle mould. When cold, these too can be buffed up in the same way to achieve a nice finish.

I hope you get good results with the above methods but if at first you don't succeed, try again until it's perfect.

Happy moulding!

Good examples of moulded wax in various shapes. (D Shannon)

Beautiful, pure beeswax show blocks. (D Shannon)

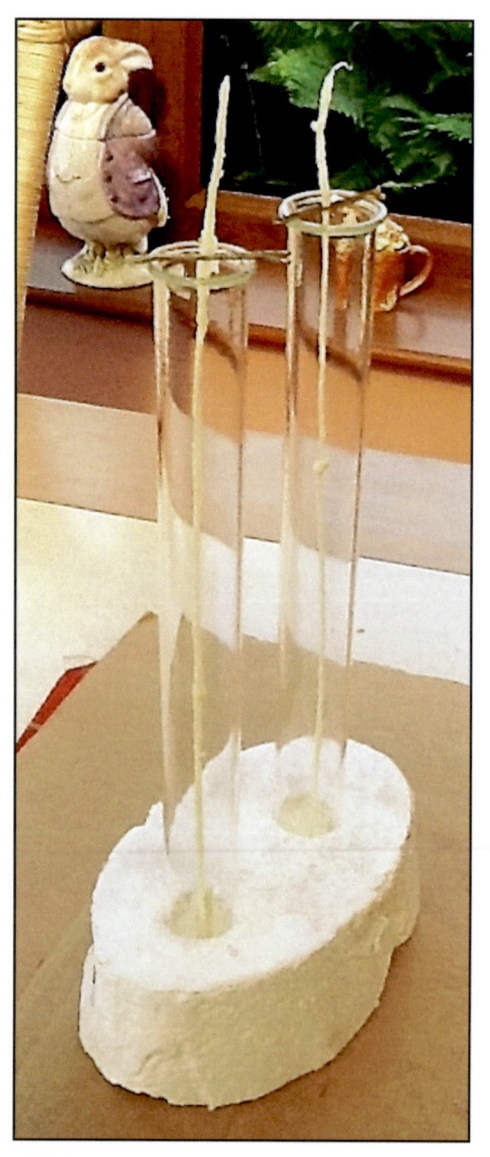

Wick primed with wax fitted carefully into the centre of a glass mould. (D Shannon)

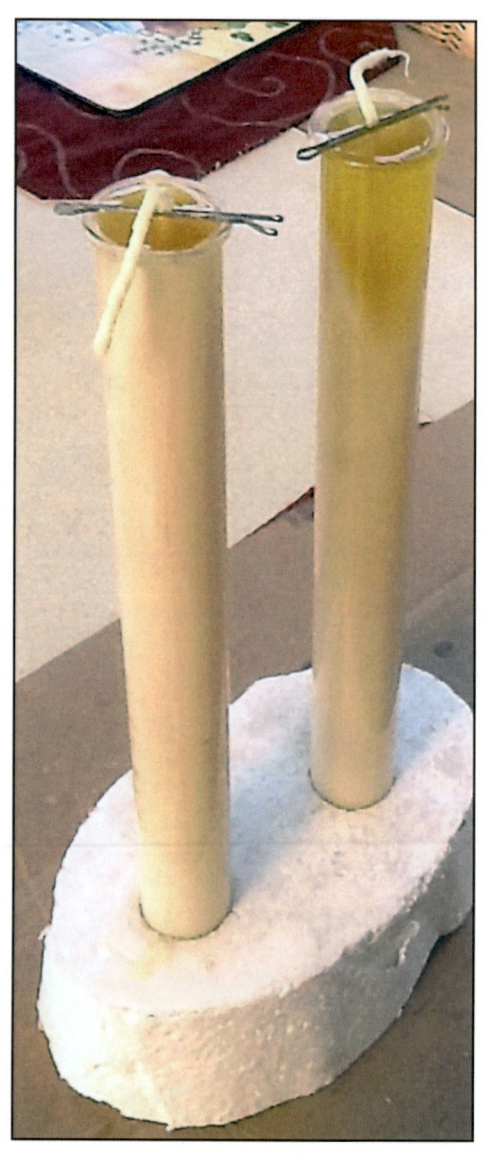

The wax must be allowed to cool very slowly. (D Shannon)

(D Shannon)

(D Shannon)

Wax exhibits can be relatively simple (D Shannon) or very complex like this steam locomotive. (John Phipps, National Honey Show, London)

Having studied this book carefully, the reader may well be on the way to winning a trophy at a local or prestigious honey show. (John Phipps, National Honey Show, London)

Lightning Source UK Ltd.
Milton Keynes UK
UKRC02n2112260916
283857UK00004B/24